THE GREAT HOME GYM HANDBOOK

WRITTEN BY
Andre Noel Potvin
Michael Jespersen

RESEARCHED BY
Michael Jespersen

EDITED BY

EXPERT: STRENGTH TRAINING & GENERAL FITNESS
Andre Noel Potvin
    MSC, CSCS, CES

EXPERT: STRETCHING & GENERAL FITNESS
Nikos Apostolopoulos
    BPHE, NCCP-L3, AACA, AAA, ACSM, IASP

GENERAL EDITOR
Michael Jespersen

COPY EDITOR
Karl Thorson

Fifth Printing

Copyright © 1999, 2003, 2006
    by Productive Fitness Products Inc.

> Consult your physician before starting any exercise program. This is of particular importance if you are over 35 and have been inactive for a period of time. The author and publisher disclaim any liability from loss, injury, or damage, personal or otherwise, resulting from the procedures in this book.

> Our thanks to Hoist for supplying the cover image of their G6 Gym. Visit their web site at **www.hoistfitness.com**

Published 1999
Productive Fitness Products Inc.
2289-135A St.
Surrey, B.C. V4A 9V2

Productive Fitness Products Inc.
1645 Jill's Court, Ste. 102
Bellingham, WA   98226

or e-mail
info@productivefitness.com

Visit our Website: www.productivefitness.com

Jespersen, Michael, 1962-
  The great home gym handbook

Includes bibliographical references.
ISBN 0-9696773-4-0

1. Exercise. 2. Home gyms. I. Potvin, André Noël, 1961- II. Thorson, Karl, 1964- III. Apostolopoulos, Nikos. IV Title
GV481.J47 1999   613.7'1   C99-910995-2

Table of Contents

Introduction	4
How To Get The Most From This Book	5
Body Diagrams	6-7
Common Myths	8-9
Home Gyms	10
Accessories & Attachments	11
Maintaining Your Home Gym	12-13
How To Set Up A Program	14-18
Aerobic Training & Equipment	19-21
Weight Training Safety Tips	22
Staying Motivated	23
Stretching	24-29

Home Gym Exercises

Chest
Vertical Bench Press	30
Lying Bench Press	31
Incline Bench Press	32
Decline Bench Press	33
Pec Flye	34
Pec Dec	35

Shoulders
Shoulder Press	36
Front Deltoid Raise	37
Standing Side Raise	38
Shoulder Shrug	39

Back
Lat Pulldown–Close Grip	40
Lat Pulldown–Wide Grip	41
Low Row	42
Mid Row	43
One Arm Row	44
Bent Over Row	45

Triceps
Tricep Pushdown	46
Dips	47
Tricep Extension	48

Biceps
Standing Bicep Curl	48
Hammer Curl	50
Preacher Curl	51

Forearms
Wrist & Reverse Wrist Curl	52

Abdominals
Crunch	53
Vertical Knee Raise	54

Legs
Leg Press	55
Leg Extension	56
Standing Leg Curl	57
Lying Leg Curl	58
Leg Abduction	59
Leg Adduction	60

Calves
Standing Calf Raise	61
Seated Calf Extension	62

Other Products	63-64

INTRODUCTION

The main reasons for purchasing a home gym are convenience and safety. The ease of working out whenever you want, without spending time driving to a fitness center, then waiting in line for machines, is what makes having your own home gym such a time-saving pleasure. Selecting the amount of resistance is as easy as pulling out the selector pin and replacing it at the desired weight, eliminating the need to lift and load heavy plates. Ensuring proper form and movement is easier since many motions are predetermined by the machine structure. For example, when doing the Pec Dec exercise, your arms move the way the machine moves, helping ensure proper form. A good machine will be adjustable to suit different sized people while still offering bio-mechanically correct exercises.

If you are really serious about fitness it makes sense to have a set up at home that is similar to a fitness center, including a multi-station gym, some free weights, a body ball and a piece of aerobic equipment.

The benefits of maintaining a strength training program include:
- *decreased body fat.* By increasing your metabolism, strength training increases the number of calories you burn, even while resting.
- *stronger bones.* Having stronger bones helps to prevent osteoporosis and other bone diseases, while preventing fractures as well.
- *improved lymphatic drainage.* Your lymphatic system is responsible for removing the toxins in your body. Weight training stimulates the movement of lymphatic fluid.
- *increased self-confidence.* Being and feeling stronger helps to relieve anxieties and fears, while giving you a certain assurance in everything you do.
- *improved body image.* Aside from seeing your muscle size and density increase, you will find yourself standing straighter with your shoulders back and head upright.
- *decreased fatigue.* Your stamina will increase and you won't tire as easily.
- *faster metabolism.* The faster your metabolism, the more efficiently your body will lose stored fat, digest and absorb dietary nutrients and burn calories taken in.
- *increased motility.* This is the speed with which food passes through your digestive system. The less time food spends sitting in your digestive tract the better.

How to get the most
FROM THIS BOOK

- ✓ From one workout to the next, strive for <u>slow and consistent increases</u> in either the amount of weight lifted or repetitions performed.
- ✓ Combine weight training and aerobic sessions together; spend 20+ minutes on aerobics and 45 minutes to one hour for weight training
- ✓ Exercise 3 to 4 times per week (both aerobics & weights)
- ✓ If you miss some workouts because of illness or vacation, don't get discouraged. Return to your regular routine as quickly as possible.
- ✓ If you feel like skipping a workout (it happens sometimes), try to do a modified workout at 50% of what you normally lift. You'll keep your workout momentum going and you'll want to get back to full strength next time.
- ✓ If you sustain a minor injury or experience discomfort while performing a particular exercise, stop the exercise, make a note in your journal listing the type of discomfort, the place where you feel it and the exercise you were doing when it happened. As soon as possible, talk to a recognized health professional and ask them to check your form.
- ✓ Perform each exercise with precision, using proper form and speed.

Body Diagrams

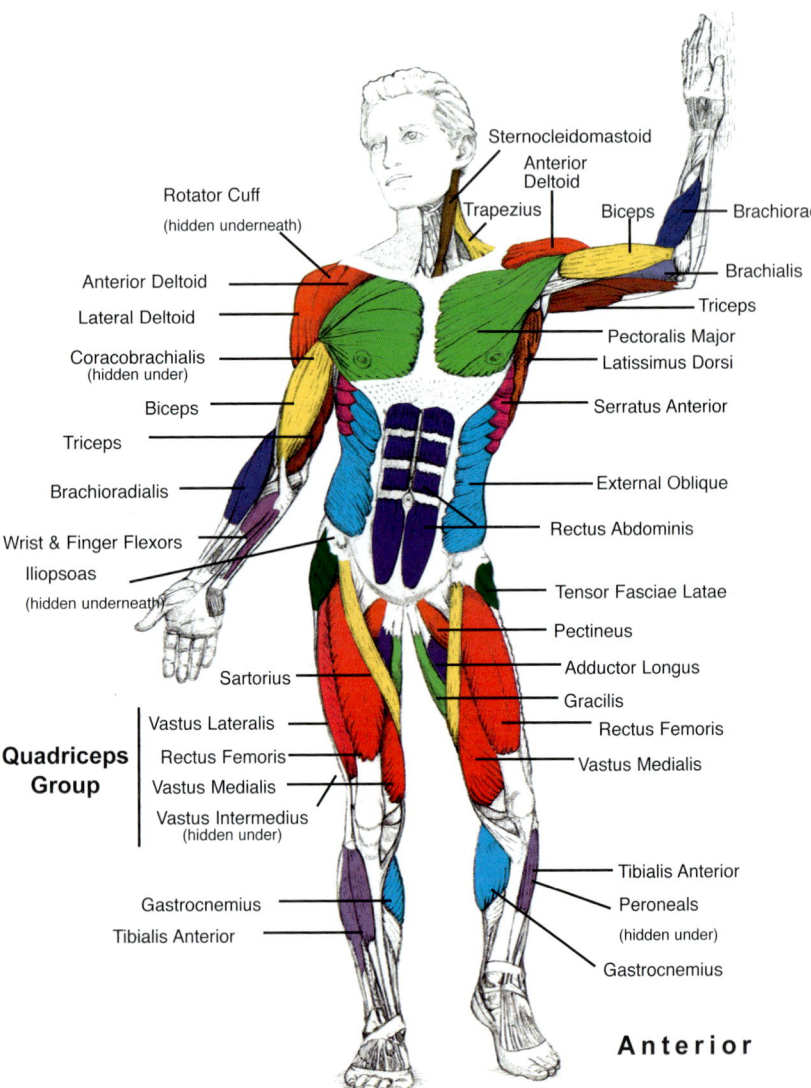

Home Gym Exercises | 7

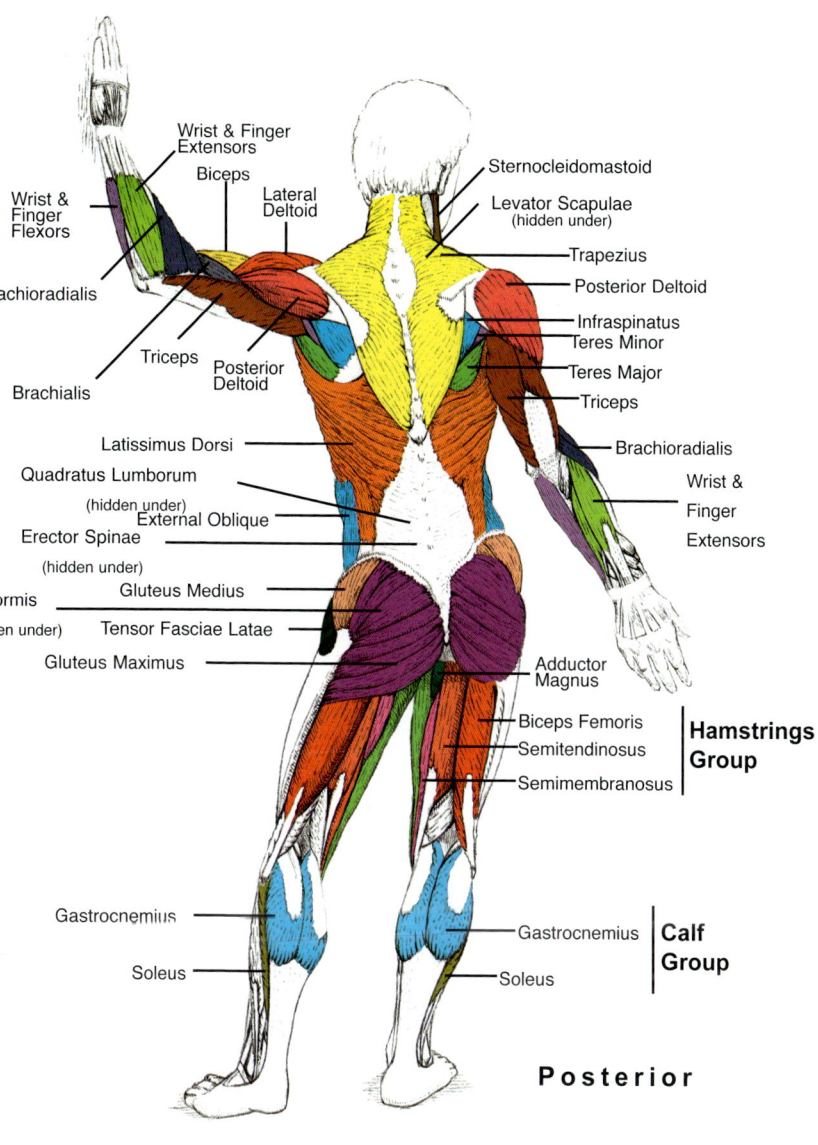

Posterior

Core Muscles= Rectus Abdominis, Erector Spinae, Internal & External Obliques

Core Stabilizers= Multifidus, Levator Ani, Transversus Abdominis

Hip Flexors= Iliopsoas, Sartorius

Hip Adductors= Adductor Brevis, Longus and Magnus, Gracilis, Pectineus

Hip Abductors= Gluteus Medius and Minimus, Tensor Fasciae Latae, Piriformis

Common MYTHS

MYTH #1: If some training is good, more is better.

Wrong. Proper rest is as important as the workout itself. Muscle and connective tissue need time to repair themselves from the stress of exercise. Overtraining can make you susceptible to injury and infection, not to mention making the realization of your goals a longer road. Allow 48 hours of rest for a muscle group before working it again.

MYTH #2: Doing hundreds of sit-ups will get rid of those extra pounds around the stomach.

This concept is called "spot reducing" and has been proven ineffective. The hope of losing a bit of the belly with countless abdominal exercises is not the best approach. One must understand that the body alone decides on which fat store it will tap to fuel the energy requirements of exercise. You can encourage the body to burn fat by engaging in prolonged exercise (more than 15 minutes) at a moderate intensity (approximately 60-65% of Max Heart rate). Where you lose fat is out of your control, therefore abdominal crunches will certainly tone the area and make it stronger but will not make a big difference in local reduction of fat.

MYTH #3: With an increase in muscularity comes a decrease in flexibility.

This brings up the matter of technique. Firstly, if you exercise a joint through a full range of motion you will maintain and possibly even increase the flexibility with that joint. Secondly, if stretching exercises are performed on a daily basis or before and after a workout, the range of motion will be maintained and almost certainly increased.

MYTH # 4: Ballistic stretching (a technique involving bouncing) Is a good way of improving flexibility.

This technique was a common practice of the past but is becoming obsolete. Ballistic-type stretching is not recommended. With this method, micro-tearing of tissue is common (resulting in muscle soreness) and the muscle-tendon unit does not maintain a stretched position long enough to be effective in permanently increasing the muscle's flexibility. The risk of injury through muscle strains outweighs the intended benefits of increased range of motion. Static stretching or a "hold for 30 to 60 seconds" technique is much more effective and safe.

MYTH #5: It's better to drink warm water than cold water to avoid muscle cramping.

Drink water that is cold (approx.. 40°F or 10°C). It has been proven that cold water is absorbed into the bloodstream faster than warmer water and has not been shown to be associated with any muscle cramps.

MYTH #6: If you stop working out all your muscle will turn to fat.

Muscle and fat are two separate and different materials and it is impossible for one to change to the other. It is possible for muscles to atrophy from disuse and at the same time fat stores increase, but this takes several months to years to accomplish.

MYTH #7: Lifting weights will build large bulky muscles in women.

If developing large muscles is a concern consider these points:

1. Testosterone, of which most women carry only a trace amount, must be present in adequate amounts for muscle growth.

2. It takes many hours of strenuous weight training to produce large muscles. Serious bodybuilders will train 4 or more hours on average each day. Some even work-out twice a day at 3-4 hours per session!

3. There may be a genetic potential for the muscle belly to develop to a large size, but most people (both male and female) do not possess this trait. Even if a woman were to have this potential, the average female testosterone levels are inadequate to build "big" muscles without some aid such as an anabolic steroid.

> **IMPORTANT**—A proper weight training program will increase a woman's strength without a significant increase in size. Most women prefer using low resistance and high repetitions in their exercise programs to define and tone their physiques.

HOME
GYMS

There are many different makes and models of home gyms on the market, ranging in features and, of course, in price and quality. Almost every home gym will have a weight stack, low pulley, middle pulley, and a high pulley.

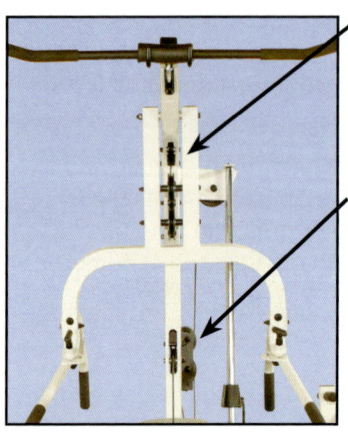

High Pulley

Exercises
- Lat Pulldown
- Tricep Pushdown

Middle Pulley

Exercises
- Abdominal Crunch
- Tricep Extension

Low Pulley

Exercises
- Leg Abduction
- Leg Adduction
- Seated Row
- Standing Bicep Curl
- Front Deltoid Raise
- Standing Side Raise
- One Arm Row
- Hammer Curl
- Wrist Curl
- Wrist Curl
- Reverse Wrist Curl

Weight Stack

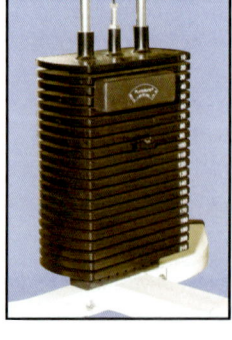

The number of weight stacks on a multi-station gym determines the number of people that can use it at the same time For instance, a two stack gym can have two people doing weight-bearing exercises at the same time.

Accessories & ATTACHMENTS

Revolving Straight Bar
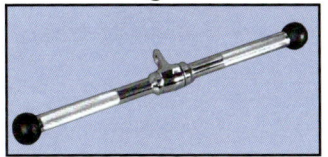
- Tricep Pushdown
- Wrist curl/reverse wrist curl
- Low Row

Seated Row/ Chinning Bar
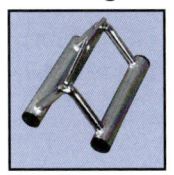
- Seated Row
- Chin ups

Revolving Multi-handle Bar
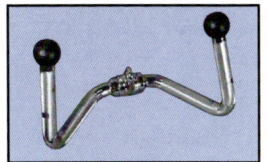
- Low Row
- Tricep Pushdown
- Lat Pulldown

Single Grip Cable Handle
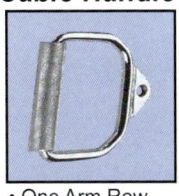
- One Arm Row
- Standing Side Raise
- Front Deltoid Raise
- Wrist Curl
- Reverse Wrist Curl

Tricep Pushdown Bar
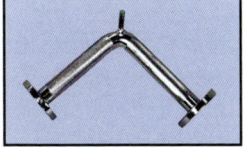
- Tricep Pushdown

Ankle Strap
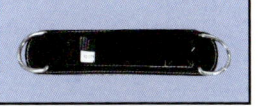
- Leg Abduction
- Leg Adduction

Tricep Rope
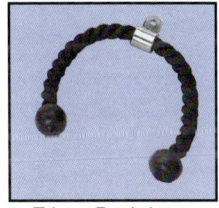
- Tricep Pushdown
- Ab Crunch
- Tricep Extension

Revolving Curl Bar
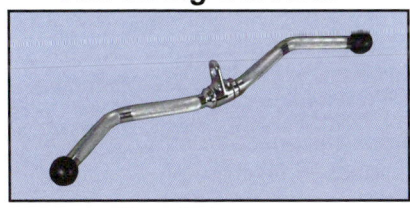
- Bicep Curl
- Lat Pulldown
- Tricep Pushdown

Revolving Lat Bar

- Lat Pulldown
- Tricep Pushdown

PlateMate
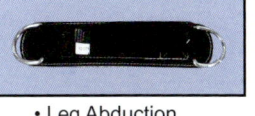

The PlateMate Brick is designed specifically for selectorized - or stack weight - machines. Build strength steadily with 2 1/2 and 5 lb increments instead of the typical 10 or 15 lb increases. When the workout's finished, your brick attaches securely to the framework of the machine, so it is always easy to find. The PlateMate is available at most fitness equipment stores or by calling PlateMate at 1-800-877-3322.

Maintaining Your
HOME GYM

By Scott Manson
Owner: Fitness FIXations (Serving Greater Vancouver Area)
E-mail: fitfix@imag.net

The good news about owning a multi-station gym is that they are relatively trouble free. That being said, there are five main areas where problems can occur: the cables, pulleys, frame, selector pins and upholstery. Do a periodic check once every 2-3 months, especially if you lift heavy weights or train often - otherwise, once every 6 months.

1. CABLES: To check your cables, pull the selector pin from the weight stack so the cable moves freely through the pulleys with no resistance. Pay particular attention to checking the areas most susceptible to wear, including the small section of cable that runs through the pulley track and also the ends of the cable where it connects to the attachments and weight stack.

What to look for:

    a) Cracks in the lamination that run across the cable.
        - a sign that plastic is becoming old and brittle.

    b) Sections of cable lamination wearing off.
        - especially in areas that come in contact with the machine, such as the pulleys.

    c) Wire poking through the cable lamination.
        - snapped strands will poke through.
        - do a visual inspection first, then run your hands lightly along the cable to feel for protruding wire.

    d) Cable or lamination has a spiral look.
        - indication that some of the wire strands have broken.

    e) Check the diameter of the cable for stretching.
        - look for areas that are wide then narrow and wide again.
        - run the cable between your thumb and forefingers to test.

Maintaining Your Home Gym | 13

If any of these conditions exist, replace the cable immediately. A snapped cable can cause injury to the user and damage to the machine. Most cables can be ordered from the dealer, or from a licensed equipment repair service where they will cut one to size.

2. POOR PULLEY CONDITION: Visually inspect the pulley for chips and cracks. Check for lateral (side to side) play. Too much could be a sign that the bearings are worn.

3. FRAME: Make sure all nuts and bolts are tight. Visually inspect the frame for stress damage and cracks (this is especially important on less expensive gyms that may have a thinner gauge steel frame).

4. SELECTOR PINS: These are normally for arm and seat adjustments and are spring loaded. Make sure they are not loose and the pin completely clears through the selection hole. (See photo below.)

5. UPHOLSTERY: Check for cracks and tears. This is mostly cosmetic, but exposed foam tends to hold water which may lead to rust damage.

> One last note about the guide rods which your weight plates move up and down along. On most machines, the guide rods do not require lubrication because there are nylon bushings in all the plates. Any lubrication you apply will eventually work its way down the guide rods and into the weight stack, causing the plates to stick together with a gum-like effect. Occasionally wipe the guide rods with a dry towel.

Selector Pin   *Pulley*

How To Set Up A PROGRAM

1. ESTABLISH GOALS

Begin by setting specific and realistic goals. Ideally, set a long-term goal and then set a series of short-term goals toward the attainment of the long term goal. Some types of goals include:

- Trim and tone physique
- Build muscle size
- Increase strength

If your goals involve losing a considerable amount of fat, then a greater aerobic component is required. It is important to focus on daily goals of constant improvement and the short and long term goals will take care of themselves.

Once you have decided on your goals, list exactly how you will attain them, including the number of workouts/week, type of activity, time of day for workout, and how you will incorporate this into your weekly schedule. Look to see what changes in your lifestyle you will have to make to accommodate the new program.

Make sure your program includes the basic components of a successful program.

PROGRAM COMPONENTS

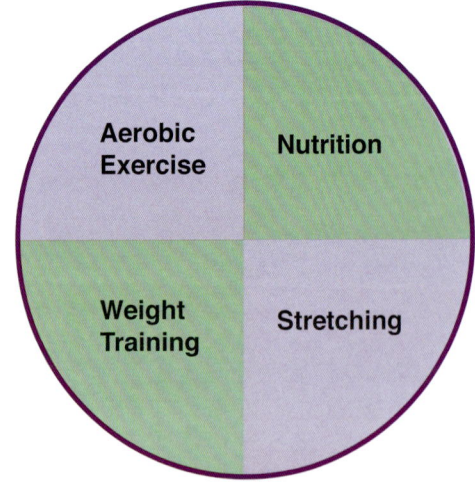

2. CHOOSE A TRAINING METHOD

The training method you select should reflect both your present fitness level and your future goals. If you are a beginner start slowly, gradually increasing the frequency and intensity of your training. A realistic schedule is both safer and easier to follow. Here are some key elements you should remember when designing your personal program:

Different training methods can be used for attaining definition, strength, or gaining muscle size (hypertrophy). A good way to decide which method is best for you, is to examine your present physique and determine your objectives. Do you prefer a moderate but well-defined body or do you want to have bulging muscles? Perhaps you are involved in a sport where rapid speed and power are required.

DEFINITION

This method of training stimulates a high cardiovascular rate and helps burn excess fatty tissue, while adding definition to the muscles. Exercises are commonly performed for 12 to 20 repetitions and 1 to 3 sets, using a light to moderate weight. Rest between sets should be from 30 to 60 seconds.

DEFINITION
REPS 12 - 20
SETS 1 - 3
REST 30 - 60 secs

HYPERTROPHY

Hypertrophy (method often used by body builders) shocks the muscles to stimulate a rapid increase in size. This is achieved by lifting moderate to heavy weight, usually 6 to 12 repetitions, and between 2 to 6 sets. Rest between sets should be kept between 1 to 1 1/2 minutes.

HYPERTROPHY
REPS 6 - 12
SETS 2 - 6
REST 1 - 1.5 mins

STRENGTH

The purpose of this method is to build muscle strength. Normally, 4 to 6 sets are performed with repetitions between 2 and 5. The weight lifted is usually very heavy, and therefore rest between sets is extended to 3 to 4 minutes.

STRENGTH
REPS 2 - 5
SETS 4 - 6
REST 3 - 4 mins

3. COMPONENTS OF A SUCCESSFUL STRENGTH TRAINING PROGRAM

- ✓ **Frequency.** Exercise each muscle group 2-3 times per week. Allow a minimum of 48 hours rest for each muscle group worked. If you are doing a total-body workout, three training sessions per week, performed on every second day, is adequate.
- ✓ **Duration.** A weight training routine should take anywhere from 45 minutes to 1 hour to complete. Add another 20 to 60 minutes if you are including aerobics.
- ✓ **Fatigue.** Try to fatigue your muscles within the suggested rep range. Fatigue is when you can't possibly do another rep without sacrificing form.
- ✓ **Range of Motion.** Moving through a complete range of motion (ROM) allows the muscle to stretch before contraction and increases the number of fibers being recruited. This produces maximum contraction and force. By working the full ROM, flexibility will be maintained or even increased.
- ✓ **Speed of Movement.** Strength training movements should be slow and controlled. Do not use momentum to complete an exercise. Momentum puts unnecessary stress on tendons, ligaments, and joints and does not develop increased strength.
- ✓ **Proper Form.** Focus on the proper motion of the exercise, while concentrating on the specific muscles being used. Do not sacrifice proper form to lift heavier weight or perform more repetitions.
- ✓ **Change Routine.** If you want to make changes in the exercises you do, wait until about the six-week point.
- ✓ **Rest Interval.** Allow a brief pause between sets to give the muscles a chance to partially recover before working them again. For hypertrophy or muscle size development allow 1 to 1.5 minutes; for endurance allow 30 to 60 seconds; and for strength allow 3-4 minutes.
- ✓ **Breathing.** Never hold your breath during any part of an exercise. Holding your breath may cause severe intra-thoracic pressure and raise blood pressure leading to dizziness, blackout or other complications. The rule of thumb is to exhale on exertion and inhale on the return part of the exercise.

4. EXERCISE ORDER

When designing a strength training routine, always try to work the larger muscle groups first. Exercises that involve more than one muscle group (compound exercises) should be at the beginning of the routine and exercises that involve only one muscle group (isolation exercises) should follow. This will prevent your muscles from becoming prematurely tired.

Order of Muscle Groups by size for;

1) Upper Body 2) Lower Body 3) Abdominal and Lower Back.

Upper Body

- Chest (pectoralis major and pectoralis minor)
- Upper back (latissimus dorsi and rhomboids)
- Shoulders (anterior, medial, and posterior deltoids and trapezius)
- Rotator cuff (supraspinatus, infraspinous, teres major and minor, and subscapularis)
- Triceps(long, medium, and short heads)
- Biceps (biceps brachii, brachialis, brachioradialis)
- Forearms (flexors and extensors)

Lower Body

- Gluteal muscle group (buttocks)
- Hip muscle group (psoas, adductors, and abductors)
- Quadriceps muscle group (vastus medialis, vastus lateralis, vastus intermedius, and rectus femoris)
- Hamstrings muscle group (semimembranosus, semitendinosus, and biceps femoris)
- Calf muscle group (soleus, gastrocnemius, anterior tibialis)

Abdominals and Lower Back

- Abdominals (transverse and rectus abdominis, and obliques)
- Lower back (quadratus lumborum and erector spinae)

REPETITION, SET, AND WORKLOAD

Repetition, also known as "reps",are the number of times an exercise is done consecutively without rest. One complete series of continuous, consecutive repetitions is called a Set. Workload refers to the amount of weight used in working a particular muscle or muscle group.

5. DESIGN YOUR ROUTINE

Step 1

Decide which of the training methods (from page 15) is best suited to accomplish your particular goals.

Step 2

Go through the exercise descriptions and select one or two exercises per body-part. Don't leave any of the body-part sections out or your routine will not be balanced. If you are trying to increase muscle size it's all right to add extra exercises to the area you want to work.

Step 3

You can either divide the exercises into upper body and lower body or keep them all together. If you only select one exercise per body part, the whole routine can be done in the same workout. If you are doing more than 12 exercises, you may want to split the routine into two or more workouts, i.e. upper body one day, lower the next day. Make sure you rest a particular muscle group 48 hours before working it again.

Step 4

Order the exercises in the workout according to the type of exercise and the muscles used (see page 17).

Step 5

Write down the exercises along with the number of repetitions and the number of sets on a piece of paper. When starting a new routine, start off with light weights and gradually increase the amount from set to set until you reach a weight you are comfortable with.

Step 6

Review and practice the components of a successful strength training program from page 16.

Sample Routine

EXERCISES	REPS	SETS	WORKLOAD
Bench Press	8-12	2 or 3	moderate
Lat Pulldown	8-12	2 or 3	moderate
Shoulder Press	8-12	2 or 3	moderate
Tricep Pushdown	8-12	2 or 3	moderate
Standing Bicep Curl	8-12	2 or 3	moderate
Leg Press	8-12	2 or 3	moderate
Leg Extension	8-12	2 or 3	moderate
Leg Curl	8-12	2 or 3	moderate
Standing Calf Raise	8-12	2 or 3	moderate
Abdominal Crunch	20-30	3	body weight

Aerobic Training
& Equipment

Aerobic exercise can be any type of activity that causes your heart rate to increase and makes you breathe harder than normal. By definition, aerobic means "with oxygen".

Fats and carbohydrates burn in the presence of oxygen to create the energy you need to continue an activity. As a fat burner, aerobic activity is particularly effective if you can maintain the activity for a minimum of 20 minutes. Put simply, the body has two sources of energy; sugar and fat. The sugar (known as glycogen) is stored in the muscles and is the easiest form of energy for your body to use. Fat requires more work to be used as energy and your body will resist using it unless the muscle runs out of glycogen.

If you are exercising at a fairly hard rate, which can be measured by taking your pulse, you should deplete the glycogen in your muscles in about 10-15 minutes. When the muscle runs low of glycogen to burn it will increase metabolizing fat as energy. This is the reason that if you are trying to burn fat you must maintain the activity for a minimum of 20 minutes. The longer you can maintain the activity without stopping the more fat you can burn. If you stop or slow down your body takes the opportunity to replace its glycogen energy stores.

The Formula for Calculating Target Heart Rate Range

Start by calculating your Maximal Heart Rate (MHR). By multiplying your MHR by upper and lower percentages you can calculate your Target Heart Rate (THR) range.

1) Calculate your approximate Maximal Heart Rate (MHR) by subtracting your age from 220. Example: 220-25 (age) =195 (MHR of a 25 year old)

2) Then to calculate your Target Heart Rate (THR) range, multiply the MHR by 55% and 85%.

Example: 195 x .55 = 107   195 x .85 = 168

Therefore the THR range for a 25 year old is between 107 and 168 beats per minute.

Periodically check your heart rate while exercising to be certain you are within the proper range. To concentrate on fat burning, keep your heart rate closer to the 65% range. To improve cardiovascular efficiency, move upward to around the 80% range.

If you are a beginner, keep your heart rate in the 55% range, gradually increasing your intensity as you become accustomed to exercise.

One minute after having completed your aerobic exercise, take your Final Heart Rate (FHR). The time it takes your body to recover from aerobic exercise will decrease as you become more fit.

> **NOTE:** You should be cautious while taking your pulse at the neck; pressure receptors in the carotid artery may slow the heart giving an artificially low reading. This method may even cause fainting in susceptible individuals; therefore, it is safest to take your pulse at the wrist.

AEROBIC EQUIPMENT

The secret to maintaining an aerobic exercise program is finding an activity you enjoy and want to stick with. Jogging, skip rope and aerobic dance are all great, inexpensive ways of aerobic training. A high quality, well constructed piece of aerobic equipment will endure many years of use. As with most things in life, you get what you pay for.

Treadmills

Treadmills are probably the most enjoyable exercise machines, whether you typically like a slow, comfortable walk or a long run. Regardless of your age or fitness level, treadmills are suited for both intense and mild training, and therefore are one of the most effective pieces of equipment.

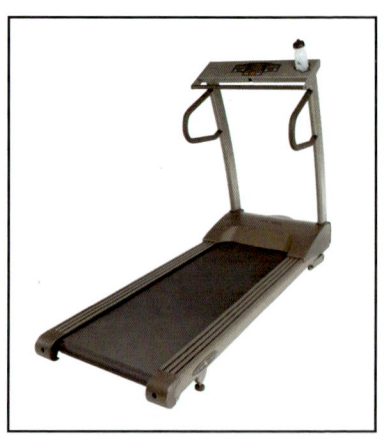

Aerobic Training & Equipment

Exercise Bike

An exercise bike focuses on the lower body, working the muscles in the legs and buttocks. Adjust the seat height so the leg has a slight bend in the knee at the pedal's lowest position.
If you find upright bikes uncomfortable try a recumbent or semi-recumbent bike.

Elliptical Trainers

An elliptical trainer is one of the most popular machines in the gym. It provides a low-impact cardio workout that protects your joints. Some elliptical trainers incorporate upper body movements as well as a lower body movements.

Rowing Machines

Rowing machines are great for working both the upper and lower body. The two types of rowing machines include the flywheel type and the hydraulic shocks type. Keep your back straight throughout the motion and don't lock your knees when your legs are extended.

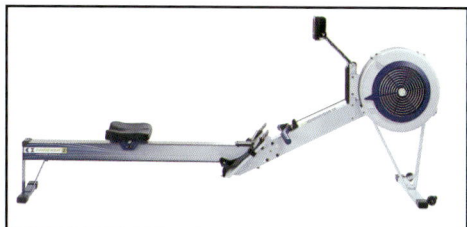

We would like to thank Vision Fitness and Concept2 for the use of their product images. Visit their web sites at **www.visionfitness.com** and **www.concept2.com**

WEIGHT TRAINING
SAFETY TIPS

✓ **Always warm up before you start a workout.**

Try to do a total body warm-up before you start training. A good example of a total body warmup is using a rowing or skiing machine. It is especially important to warm up the specific muscle groups you are going to be using. This can be as simple as performing a warm-up set of high repetitions and light weight (25% to 50% of your training weight) for each exercise.

✓ **Use proper posture.**

Maintaining proper posture will greatly reduce chances of injury and maximize exercise benefit. When standing always keep your feet shoulder-width apart. Do not lock your knees: it puts an unnecessary strain on them. Keep your back flat and straight, making sure not to twist or arch it in order to complete an exercise.

✓ **Use proper form.**

Focus on only working the muscle groups intended for the exercise you are doing. If you feel strain elsewhere you may need someone to critique your exercise motion or reevaluate the amount of weight you are lifting. Keeping proper form also means lifting in a smooth fluid motion. Know when your muscles are too tired to keep going.

✓ **Breathe properly.**

Never hold your breath during any part of an exercise. Holding your breath may cause severe intra-thoracic pressure and raise blood pressure leading to dizziness, blackout or worse! The rule of thumb is to exhale slowly on exertion and inhale on the return part of the exercise.

✓ **Stop training if you feel pain.**

If you feel pain during a specific exercise stop immediately. Any continuation may aggravate an existing injury. Reevaluate your routine to make sure you are doing a proper warm-up. Decrease the amount of weight you are lifting. Talk to a recognized health professional.

Staying Motivated

✓ **Set short-term goals to achieve long-term goals.**

After setting long-term goals go back and break them up into smaller goals that are easier to achieve. Reward yourself each time you accomplish a goal.

✓ **Make a point of having fun during your workouts.**

Get a training partner: someone who can offer support and encouragement when working out. Buy a walkman and exercise to your favorite music. Workouts are so much easier when they are fun.

✓ **Establish a definite time and place to work out.**

Force yourself to be consistent in your workouts for the first couple of months. This will help you to feel you are accomplishing your goals, it will set a pattern for future workouts, and it will help you to make exercise a part of your lifestyle.

✓ **Keep a training journal to monitor your progress.**

If you're truly dedicated to getting the most out of your exercise routine you should consider keeping an accurate record of your weight training sessions. The whole benefit of doing any type of weight training comes in the form of progressive resistance. By gradually increasing either the weight lifted, the number of repetitions, or the number of sets, you challenge your body to perform a little better each time. The direct benefits of keeping a journal are: more organized routines, knowing how and when to challenge your body, reduced chance of injury by over-lifting, a sense of accomplishment from actually seeing your progress on paper.

STRETCHING

BY ANDRE NOEL POTVIN, MSc, CSCS, CES

Why Stretch?

Regular stretching helps maintain and improve flexibility. The definition of flexibility is a joint's ability to move through a normal range of motion (ROM). Each joint has its own degree of flexibility; therefore, it's possible to be very flexible in one joint and stiff in another. The primary limitation in joint ROM is due to the tough connective tissue running through the muscle belly. Other factors that influence flexibility include:

- age
- genetics
- activity (previous exercise experience)
- joint structure (injury or no injury)
- gender (women are generally more flexible than men)
- body temperature (slightly warmer than normal is more effective)
- opposing muscle tightness (opposing muscles are responsible for returning limbs to their original position).

Stretching is the practice of tissue elongation, or lengthening muscle and connective tissue for the purpose of reducing tension around a specific joint. Stretching allows the joint to move more freely. Some benefits of stretching include:

- increased joint range of motion
- reduced joint stress due to muscular imbalances
- reduced chronic soft-tissue pain (i.e., neck, back, knees, etc.)
- increased relaxation
- enhanced well-being

When stretching, keep the following points in mind.

- Stretch to a mild-intensity (30%-40% of maximum intensity). The stretch should feel like a comfortable pull.
- Hold stretches for 30-60 seconds, until the muscle relaxes. When you begin a stretch, your muscles will feel tight; this feeling subsides as the muscle relaxes, then elongates.
- Stretch when your muscles are warm, ideally after physical activity, such as resistance training or aerobics. Stretching with warm muscles enhances results. Avoid stretching cold muscles.
- Pay extra attention to your tightest joints. Flexibility is joint-specific; focus on all joints with restricted ROM.
- Proper body alignment is critical for getting maximum results. Carefully study and follow the stretch positions and explanations in this handbook.
- Repeat each stretch 2-3 times.
- Breathe deeply as you stretch; this enhances relaxation by stimulating the Central Nervous System (CNS).

Home Gym Exercises | 25

IMPORTANT SAFETY TIPS
- STOP stretching if you feel pain.
- NEVER push against or force a joint beyond its limit.
- NEVER hold a stretch longer than 90 seconds. Doing so could weaken the tissue and increase the risk of injury and/or irritation.
- If you feel pain during any of these stretches, STOP IMMEDIATELY and see your physician.

Flexibility Training Guidelines

Intensity	• Using a scale of 1-10, stretch at about a 3-4 intensity level (1=very mild stretch, and 10=extreme stretch). You should feel a comfortable pulling sensation, never pain. 1 out of 10=very very mild 10 out of 10=extremely intense
Time	• Hold each stretch for 30-60 seconds. • Perform each stretch 2-3 times.
Other Variables	• For optimal results, stretch after a warm-up or aerobic activity when the muscles are warm.

Stretch Routine

 Neck Retractions/Chin Tucks
(Stretches: neck extensors)

- In a standing position, poke your chin and head forward, then draw your chin backward, flattening the back of your neck.
- Keeping neck retracted (chin in), tuck chin down, toward your chest.
- Hold for 30-60 seconds; repeat 2-3 times. Repeat often during the day.

 Head Tilts
(Stretches: scalenes, upper trapezius)

- Tilt head to the right and lower your left shoulder.
- Place right hand on left side of head to gently intensify stretch.
- Hold for 30-60 seconds; repeat 2-3 times. Switch sides.

Caution: Be very gentle when intensifying this stretch.

The Great Home Gym Handbook

3. Head Turn
(Stretches: neck rotators)

- Place right index and middle finger on left side of jaw.
- Place left hand on back of head, on the right side.
- Gently rotate head to the right, using hands to intensify stretch.
- Hold for 30-60 seconds; repeat 2-3 times. Switch sides.

4. Ball Arch
(Stretches: chest, ribs, shoulders, abs, spine)

- Lie face-up on a stability ball.
- Place hands behind head, holding abs tight.
- Squeeze shoulder blades together, opening elbows to the side.
- Arch your back over the ball with feet flat on floor, keeping neck neutral as you slightly look up toward ceiling.
- Breathe deeply, expanding the chest.
- Start by holding for 5 seconds, then releasing. Gradually increase over time to hold for 30-60 seconds; repeat 2-3 times.

5. Handcuff Towel
(Stretches: anterior deltoid, chest, biceps)

- Hold a towel behind your back with palms facing body.
- Squeeze shoulder blades together and pull arms backward.
- Keep abs tight and don't arch lower back.
- Hold for 30-60 seconds; repeat 2-3 times.

Note: Avoid leaning forward, rolling shoulders forward or poking your neck forward.

Advanced: Grip fingers together and repeat as above without the towel.

6. Overhead Reach
(Stretches: latissimus dorsi)

- Standing or kneeling, interlace fingers and reach arms overhead, palms down.
- Keep neck neutral; avoid poking head forward.
- Hold for 30-60 seconds; repeat 2-3 times.

Variation: Clasp the left wrist, pulling arm up and to the right. Repeat on left.

Variation

Home Gym Exercises | **27**

Seated Arm Cross-Over Hug
(Stretches: rhomboids, middle and lower trapezius, erector spinae)

- Sit with legs slightly bent in front of you.
- Cross arms, keeping them straight; hold right thigh with left hand and left thigh with right hand.
- Sit back, using your abs.
- Hold for 30-60 seconds; repeat 2-3 times.

Shoulder Towel Stretch
(Bottom arm stretch– external rotators, anterior deltoid)
(Top arm stretch - posterior deltoid, triceps)

- Hold one end of towel in right hand, raising right arm overhead.
- With towel hanging behind back, grab other end with left hand.
- Pull up on towel with right hand, straightening arm, to stretch left (bottom) shoulder.
- Pull down on towel with left hand, straightening arm, to stretch right (top) shoulder.
- Hold for 30-60 seconds; repeat 2-3 times. Switch sides.

Note: Keep shoulder blades squeezed together.

Bottom arm stretch   Top arm stretch

Single-Knee Corkscrew
(Stretches: gluteus maximus, obliques, erector spinae, piriformis)

- Sit with left leg straight and right leg bent, knee close to chest and right foot on opposite side of left knee.
- Place right hand on floor behind you for support.
- Wrap left arm around right knee, pulling up into left shoulder.
- Rotate torso until you feel a comfortable stretch.
- Breathe deeply. As you exhale, twist a little more.
- Hold for 30-60 seconds; repeat 2-3 times. Switch sides.

Note: This stretch works best when you sit up as straight as possible.

10. Runner's Hip Stretch

(Stretches: tensor fasciae latae, iliopsoas, rectus femoris, obliques, erector spinae, spine)

- Take a large step forward with right leg.
- Place a stability ball under right buttock for support, keeping most of your weight on legs.
- Bend left knee down and toward front leg.
- Twist torso to the right, placing right hand on ball (or right buttock), and left hand on right side of right thigh.
- Squeeze buttocks and tilt pelvis forward. (Imagine pelvis is a bucket tilting to pour water behind you.)
- You should feel the stretch in the left hip and thigh.

Hold for 30-60 seconds; repeat 2-3 times. Switch sides.

Note: Avoid arching lower back

Advanced: Cross left foot behind you and to the right; repeat as above.

Variation

11. Seated Towel Stretch

(Stretches: hamstrings, gastrocnemius)

- Sit with both legs slightly bent in front of you.
- Wrap a towel or rope around left foot.
- Gently straighten left leg until you feel a comfortable stretch.
- Pull left toes and foot toward you.

Hold for 30-60 seconds; repeat 2-3 times. Switch sides.

Note: Lift chest and straighten back by arching through lower back; retract shoulder blades (avoid rounding them forward) and keep neck neutral, or look slightly down.

12. Bent-Knee Stretch

(Stretches: quadriceps, hip flexors)

- Hold left ankle while standing.
- Pull left heel to buttocks, or until you reach a comfortable stretch.
- Point left knee toward floor.
- Tilt pelvis. (Imagine pelvis is a bucket tilting to pour water behind you.)
- Keep spine straight and upright, chest lifted and head neutral.

Hold for 30-60 seconds; repeat 2-3 times. Switch sides.

Modification: For a stiff knee or quadriceps, place left foot on a bench or chair.

Variation

Home Gym Exercises | **29**

⑬ Cross-Leg Stretch
(Stretches: gluteus medius and minimus)

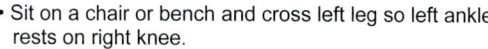

- Sit on a chair or bench and cross left leg so left ankle rests on right knee.
- Sit straight with chest lifted and shoulder blades slightly squeezed together.
- Gently press on left knee with left hand until you feel a comfortable stretch.
- Hold for 30-60 seconds; repeat 2-3 times. Switch sides.

⑭ Inner Thigh Stretch
(Stretches: adductors)

- Stand with feet wide apart, both hands on left thigh.
- Perform a slight squat, leading with buttocks and keeping chest lifted.
- Shift your weight to the left until you feel a comfortable stretch.
- Angle left foot a little toward the left, with left knee aligned over second toe.
- Point right foot forward, keeping sole flat on floor.
- Hold for 30-60 seconds; repeat 2-3 times. Switch sides.

⑮ Bent-Knee Calf Stretch
(Stretches: soleus)

- Step forward with right foot, placing right heel on floor, toes up and both hands on left thigh.
- Slowly squat with left leg until you feel a comfortable stretch in left calf. Keep left heel on floor.
- Pull left toes toward your shin to intensify.
- Hold for 30-60 seconds; repeat 2-3 times. Switch sides.

Note: Keep heels down and weight on back leg; lift chest and slightly arch lower back.

⑯ Straight-knee Calf Stretch
(Stretches: gastrocnemius)

- Step forward with right foot.
- Straighten left leg, gradually pressing left heel to floor.
- Stop when you feel a comfortable stretch in left calf.
- To intensify, pull left toes toward shin.
- Hold for 30-60 seconds; repeat 2-3 times. Switch sides.

Note: Keep your heels down and weight on back leg; lift chest and slightly arch lower back.

Chest

HOME GYM
EXERCISES

Vertical Bench Press

Muscles Worked
Pectoralis Major, Anterior Deltoid, Triceps

START　　　FINISH

Using the lower handles incorporates more tricep and deltoid muscles.

1. Adjust the seat height so that your elbows are at a 90-degree angle with your arms by your sides in the start position.
2. Sit comfortably in the seat with your back flat against the back rest.
3. Slowly begin to push the handles forward until your arms are almost completely straight. Keep a slight bend at the elbows so as not to lock your arms in the extended position.
4. Pause for a moment, then slowly return toward the start position.
5. Keep a constant tension on the weights. Don't let the metal plates rest on the stack between repetitions.
6. Repeat the exercise in a slow and controlled fashion until you have completed the set.

Lying Bench Press

Muscles Worked: Pectoralis Major, Anterior Deltoid, Triceps

Chest

1. Adjust the position of the bench so that you are starting this exercise with hands above the mid-chest line and elbows bench level.
2. Slowly begin to push upward until your arms are almost completely straight. Keep a slight bend at the elbows so as not to lock your arms in the extended position.
3. Pause for a moment, then slowly return toward the start position.
4. Avoid dropping your elbows below bench level.
5. Repeat the exercise in a slow and controlled fashion until you have completed the set.

Breathe Properly. Exhale on exertion and inhale on return.

Incline Bench Press

Muscles Worked: Pectoralis Major, Anterior Deltoid, Triceps

1. Adjust the seat angle so that the bench is approximately at a 30 to 45-degree incline. Position the bench so that your hands are in line with your upper chest.
2. Sit comfortably in the bench with your back flat against the back rest.
3. Slowly begin to push the handles upward until your arms are almost completely straight. Keep a slight bend at the elbows so as not to lock your arms in the extended position.
4. Pause for a moment, then slowly return toward the start position.
5. Keep a constant tension on the weights not letting the metal plates rest on the stack between repetitions.
6. Repeat the exercise in a slow and controlled fashion until you have completed the set.

Breathe Properly. Exhale on exertion and inhale on return.

Decline Bench Press

Muscles Worked: Pectoralis Major, Anterior Deltoid, Triceps

1. Adjust the seat angle so that the upper and lower bench are approximately at a 20-degree decline. Position the bench so that your hands are in line with your mid chest.
2. Lie on the bench with your back flat against the back rest and feet supported to prevent the back from arching.
3. Slowly begin to push the handles up until your arms are almost completely straight. Keep a slight bend at the elbows so as not to lock your arms in the extended position.
4. Pause for a moment, then slowly return toward the start position.
5. Keep a constant tension on the weights. Don't let the metal plates rest on the stack between repetitions.
6. Repeat the exercise in a slow and controlled fashion until you have completed the set.

> The return phase of an exercise is as important as the execution. The return should be slow and controlled.

Pec Flye

Muscles Worked

Pectoralis Major, Anteriod Deltoid

1. Adjust the seat height so that your arms are close to shoulder height in the start position.
2. If the upper mechanism of the arm has an adjustment, move the locking pin to the point where your elbows are in line with the shoulders, and not behind the shoulders.
3. Sit comfortably in the seat with your back flat against the back rest.
4. Slowly bring both your arms forward in a semi circular fashion until your hands meet in the middle: motion resembles hugging a tree. Keep your elbows fixed at shoulder level, making sure not to drop them.
5. Pause for a moment, then slowly return toward the start position.
6. Keep a constant tension on the weights. Don't let the metal plates rest on the stack between repetitions.
7. Repeat the exercise in a slow and controlled fashion until you have completed the set.

The return phase of an exercise is as important as the execution. The return should be slow and controlled.

Home Gym Exercises | **35**

Pec Dec

Muscles Worked
Pectoralis Major, Anteriod Deltoid

Chest

1. Adjust the seat height so that your upper arms are about shoulder level in the start position.
2. If there is an arm adjustment mechanism, move the locking pin to the point where your elbows are in line with the shoulders, and not behind the shoulders.
3. Sit comfortably in the seat, with your back flat against the back rest.
4. Slowly bring both your arms forward in a semi circular fashion until your hands meet in the middle. Keep your upper arms parallel to the floor.
5. Pause for a moment, then slowly return toward the start position.
6. Keep a constant tension on the weights. Don't let the metal plates rest on the stack between repetitions.
7. Repeat the exercise in a slow and controlled fashion until you have completed the set.

Shoulder Press

Muscles Worked

Anteriod Deltoid, Middle Deltoid, Triceps, Trapezius, Posterior Deltoid

1. Depending on the type of multi-station gym, the set up for this exercise varies widely. However, the basic fundamentals for the Shoulder Press are the same.
2. Sit comfortably in the seat with your back supported against the back rest.
3. Adjust the height of the handles (or seat) so your hands are about ear level.
4. Push the handles straight up. Keep a slight bend at the elbows so as not to lock your arms in the extended position.
5. Pause in the upper position for a moment and then slowly lower.
6. Keep a constant tension on the weights. Don't let the metal plates rest on the stack between repetitions.
7. Repeat the exercise in a slow and controlled fashion until you have completed the set.

Home Gym Exercises | **37**

Front Deltoid Raise

Muscles Worked

Anteriod Deltoid, Coracobrachialis

Shoulders

Two Arm

One Arm

1. Stand with your feet spaced comfortably apart.
2. Bend your knees slightly.
3. Grip the handle with either one or two hands, palms toward you.
3. Ideally, the first time you try this exercise use both arms and the minimum weight possible. Then gradually increase the weight until you find the correct resistance level.
4. Keeping your arm(s) straight, raise the handle directly in front of you until it reaches shoulder level.
5. Move in a slow and controlled fashion.
6. Hold in the upper position for a moment, then slowly lower back down.
7. Repeat until the set is completed.

Standing Side Raise

Muscles Worked

Anteriod Deltoid, Middle Deltoid, Posterior Deltoid, Trapezius

Shoulders

1. Stand with your feet comfortably spaced apart.
2. Bend your knees slightly.
3. Ideally, the first time you try this exercise use the minimum weight possible. Then gradually increase the weight until you find the correct resistance level.
4. Grip the handle with your right hand, palm facing toward your left thigh.
5. Keeping your arm straight, elbow slightly bent, raise the handle, bringing it sideways across your body in an upward diagonal motion until it reaches shoulder level.
6. Move in a slow and controlled fashion.
7. Hold in the upper position for a moment, then slowly lower back down.
8. Repeat until the set is completed.

Shoulder Shrug

Muscles Worked
Upper Trapezius, Levator Scapulae

START

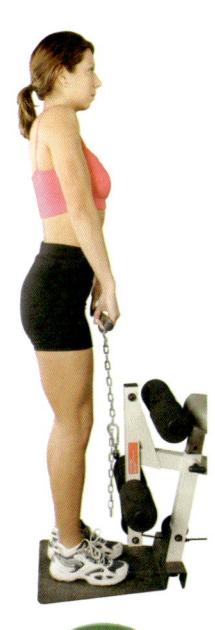
FINISH

Back View

1. Stand upright, knees slightly bent and your feet comfortably spaced apart.
2. Grip the handles with your palms facing inward and your arms completely vertical and in line with your body.
3. Slowly raise your shoulders toward your ears.
4. Keep your arms straight and do not arch your back in an effort to lift higher.
5. Move in a slow and controlled fashion.
6. Hold in the upper position for a moment, then slowly lower back down.
7. There should be constant tension as you raise and lower your shoulders.

The return phase of an exercise is as important as the execution. The return should be slow and controlled.

Lat Pulldown Close Grip

Muscles Worked

Latissimus Dorsi, Biceps, Rhomboids, Posterior Deltoid, Teres Major, Infraspinatus

START

FINISH

1. Straddle the seat so that you are facing the machine.
2. Grip the bar with your hands slightly closer than shoulder width apart, your palms facing toward you, and your fingers over the top and thumb under. Your elbows should be slightly bent.
3. Lean back slightly and slowly pull the bar toward your upper chest.
4. Keep your body in a fixed position, moving only your arms and shoulders.
5. Pause when the bar reaches your upper chest.
6. Slowly let the bar return to the start position, resisting the pull throughout the range of motion.

Variations:
Underhanded Grip—incorporates more bicep muscles.
Wide Grip—incorporates more of the smaller muscles such as teres major, infraspinatus, and posterior deltoid
Narrow Grip—incorporates more latissimus dorsi.

Lat Pulldown Wide Grip

Muscles Worked

Latissimus Dorsi, Biceps, Rhomboids, Posterior Deltoid, Teres Major, Infraspinatus

START — FINISH

1. Straddle the seat so that you are facing the machine.
2. Grip the bar with your hands wider than shoulder width apart, your palms facing away from you, and your fingers over the top and thumb under. Your elbows should be slightly bent.
3. Lean back slightly and slowly pull the bar toward your upper chest.
4. Keep your body in a fixed position, moving only your arms and shoulders.
5. Pause when the bar reaches your upper chest.
6. Slowly let the bar return to the start position, resisting the pull throughout the range of motion.

Variations:

Underhanded Grip—incorporates more bicep muscles.

Wide Grip—incorporates more of the smaller muscles such as teres major, infraspinatus, and posterior deltoid

Narrow Grip—incorporates more latissimus dorsi

Low Row

Muscles Worked

Latissimus Dorsi, Trapezius, Rhomboids, Posterior Deltoid, Biceps, Erector Spinae

1. Sitting on either the floor or a low row station, brace your feet against the foot supports.
2. Lean forward and grip the handle with both hands, your palms facing upward. Your knees should be slightly bent.
3. Slowly begin to pull the bar toward your mid-abdominal region (belly button). Your legs and upper body should be in a fixed position, with only your arms and shoulders moving.
4. Pause when the bar reaches your abs and slowly let the bar return to the start position, resisting the pull throughout the range of motion.
5. Your arms should never be pulled completely straight: always keep a slight bend in the elbow.

The return phase of an exercise is as important as the execution. The return should be slow and controlled.

Home Gym Exercises | **43**

Mid Row

Muscles Worked

Latissimus Dorsi, Trapezius, Rhomboids, Biceps, Erector

START   FINISH

1. Not all multi-station gyms can be adjusted to allow for this exercise.
2. Straddle the seat so that you are facing the machine.
3. Set up the machine so your chest is braced against the back rest and your arms are extended to grip the handle. Your elbows should be slightly bent.
4. Begin slowly pulling the bar toward your upper chest.
5. Keep your body in a fixed position, moving only your arms and shoulders.
6. Pause when the bar reaches your upper chest.
7. Slowly let the bar return to the start position, resisting the pull throughout the range of motion.

The return phase of an exercise is as important as the execution. The return should be slow and controlled.

One Arm Row

Muscles Worked

Latissimus Dorsi, Trapezius, Rhomboids, Posterior Deltoid, Biceps, Erector Spinae

START   FINISH

1. Stand with your feet spaced 3-4 feet apart, left foot in front of the right.
2. The leg in front should only have a slight bend, whereas the leg behind is bent considerably more. Lean forward slightly and support yourself with your left hand on the left thigh.
3. Grip the handle and position yourself so that your arm and the cable are at a 45-degree angle.
4. Keep your head neck and back in line and in a fixed position.
5. Begin pulling the handle toward your right hip in a slow and controlled fashion until your hand is next to the hip.
6. Pause, then slowly release the tension until your arm is back to the starting position.
7. After completing a set on one side switch to the other: (with right foot in front of the left behind).

Home Gym Exercises | **45**

Bent Over Row

Muscles Worked

Latissimus Dorsi, Trapezius, Rhomboids, Posterior Deltoid, Biceps, Erector Spinae

START　　　FINISH

1. Place a seated row/chinning bar attachment on the end of the cable.
2. Stand facing the low pulley, with your feet spaced comfortably apart
3. Bend your knees slightly.
4. Keeping your back straight, lean forward until your arms and the cable are at a 45 degree angle.
5. Keep your head, neck and back in line and in a fixed position.
6. Begin pulling the handle toward your lower rib cage in a slow and controlled fashion.
7. Pause when the handle reaches your body, then slowly release the tension until your arms are back in the start position.
8. Repeat the exercise until you complete the set.

Tricep Pushdown

Muscles Worked: Triceps

START

FINISH

1. Stand upright, knees slightly bent and your feet comfortably spaced apart.
2. Grip the bar with your fingers curled over top, hands spaced about a foot apart, and palms facing down.
3. Pull the bar down until your elbows are at your sides and your upper and lower arms form a "V" shape. This is the start position.
4. Keep your elbows fixed to your sides and push the bar down until your arms are straight and the bar nearly touches the upper thigh.
5. Only your forearm should move, the upper arm and elbow remain fixed.
6. Pause for a moment, then slowly return toward the start position, resisting the pull all the way back to the start position.
7. Repeat exercise in a slow and controlled fashion until you have completed the set.

Home Gym Exercises | 47

Dips

Muscles Worked
Triceps, Pectoralis Major, Anteriod Deltoid

START   FINISH

Triceps

1. This exercise requires significant upper body strength.
2. Start with your arms straight and hands gripping the parallel handles. All of your weight is suspended on the handles.
3. Slowly lower yourself, bending at the elbows, until you feel your upper arms and shoulders start to tighten.
4. Pause at this point then slowly push back to the start position.
5. Do not lock out your elbows in the upper position, keeping a constant tension on the tricep muscles.

Variations:
Using a platform, step up into position then slowly lower your body down. Vary the resistance using your legs.

> The return phase of an exercise is as important as the execution. The return should be slow and controlled.

Tricep Extension

Muscles Worked: Triceps

1. Attach the tricep rope to the middle pulley.
2. Grip the handles with your palms facing each other and your hands above your head so that the cable clears the top of your head.
3. Your upper arm (shoulder to elbow) should remain fixed in place throughout the exercise.
4. With a constant tension on the cable, pull the rope forward until your arms are almost straight.
5. Pause, then slowly let your lower arms return to the start position.
6. Make sure the motion is completely controlled.

Breathe properly. Exhale on exertion and inhale on return.

Home Gym Exercises | **49**

Standing Bicep Curl

Muscles Worked: Biceps, Brachialis

START

FINISH

Biceps

1. Attach the revolving straight bar to the low pulley.
2. Stand upright, knees slightly bent, your feet comfortably apart.
3. With the straight bar directly in front of you, palms facing up, begin to lift up toward your shoulders, keeping your elbows fixed.
4. If you are arching your back or jerking the bar in an effort to complete the exercise, use a lighter weight.
5. Once in the upper position hold for a second, then slowly lower the bar back down until your arms are fully extended downward.

> **The return phase of an exercise is as important as the execution. The return should be slow and controlled.**

Hammer Curl

Muscles Worked: Brachioradialis, Biceps

Biceps

START

FINISH

1. Attach the tricep rope to the low pulley (similar to standing bicep curl).
2. Stand with your feet shoulder width apart, knees slightly bent and palms toward each other.
3. Keeping your elbows in a fixed position, lift the handles straight up until your hands almost touch your chest. The palms face each other throughout the exercise.
4. Pause at the top then slowly lower the handles back to the start position.
5. Make sure the motion is slow and controlled; do not use momentum to perform this exercise.

Preacher Curl

Muscles Worked: Brachialis, Biceps

1. Sit facing the machine with your upper arms resting flat on the preacher curl pad.
2. Grasp the handle with an underhand grip, about a foot apart, and your arms fully extended. Tighten your abdominals and set your back.
3. Slowly curl the bar up to the front of your shoulders while keeping your upper arms still.
4. Pause briefly in the finish position, then slowly lower the bar back to the start position.

Variations:
Narrow Grip—increases load on the outer bicep.
Wide Grip—increases load on the inner bicep

Wrist Curl & Reverse

Muscles Worked

Wrist Flexors, Wrist Extensors, Finger Flexors

For Reverse: Have the palms facing <u>downward</u>.

Forearms

1. Sit comfortably on the seat with your feet spread shoulder-width apart.
2. Grip a straight bar so that your palms are facing up.
3. Support your forearms by placing them on your thighs so that the wrist joint is at the edge of your knee and your palm is facing upward.
4. Relax your wrists so that the bar is in the lower position. Hold the bar at the end of your curled fingers.
5. Bring the bar upward with your hand until your wrist is completely flexed.
6. Your elbow and forearm should remain in contact with your thigh throughout the exercise.
7. Slowly lower the bar back to the starting position.

Home Gym Exercises

Crunch

Muscles Worked
Rectus abdominis

START

FINISH

1. Sitting in the seat, grip the tricep rope on either side of your neck pulling it tight against the back of your neck.
2. Make sure your buttocks are firmly at the back of the seat.
3. Slowly begin to curl forward from the waist, bringing your chest toward your thighs. If possible, lock your feet behind the foot rollers to stabilize your lower body.
4. Make a conscious effort to use your abdominal muscles in performing this exercise. Do not bend at the hip.
5. Move in a slow and controlled fashion.
6. Pause at the lower position for a moment, then slowly return to the start position.
7. Repeat until the set is completed

Breathe properly. Exhale on exertion and inhale on return.

Abdominals

Vertical Knee Raise

Muscles Worked

Hip Flexors, Rectus abdominis

START

FINISH

Abdominals

1. Standing on the support platform, get into position in the VKR chair, with your back flat against the rest while gripping the handles for stability.
2. Begin slowly lifting your feet off the platform until all your weight is suspended on your forearms.
3. With a slow and consistent motion, bring your knees up toward your chest.
4. Once in the upper position, hold for one second, then slowly lower legs back down.
5. Keep your upper body tight with the shoulders fixed. Do not let your body sink down through the shoulders.

> The return phase of an exercise is as important as the execution. The return should be slow and controlled.

Home Gym Exercises

Muscles Worked

Quadriceps Group, Gluteus Maximus, Hamstrings Group

1. Adjust the seat position to accommodate your leg length. Your knee joints should be at a 90-degree angle in the start position.
2. Place your feet about shoulder-width apart on the foot pad.
3. Ensure your lower back is supported by the back rest.
4. Grip the side handles for stabilization as you slowly push the foot pad out. Keep your knees parallel making sure they don't collapse inward.
5. Be sure not to lock your knees out when you have fully extended your legs.
6. Pause in the extended position, then slowly let the foot pad return to its starting position. Be careful not to let the metal weight plates slam back together.
7. Repeat the exercise in a slow and controlled fashion until you have completed the set.

Leg Extension

Muscles Worked
Quadriceps Group

1. Sit either with your back against the bench, or in an upright position if your bench lacks a support.
2. Adjust the seat position so that the backs of your knees are against the roller or seat edge.
3. Grip the side handles to help stabilize your body while performing this exercise. Place the front of your ankles behind the lower foam rollers.
4. Slowly straighten your legs until they are completely extended.
5. Pause briefly then return to the start position in a slow and controlled fashion. Do not let your knees come back beyond the 90-degree position on the return phase.
6. Keep a constant tension on the weights without letting the metal plates rest on the stack between repetitions.
7. Repeat exercise in a slow and controlled fashion until you have completed the set.

The return phase of an exercise is as important as the execution. The return should be slow and controlled.

Home Gym Exercises | **57**

Standing Leg Curl

Muscles Worked
Hamstrings Group

START   FINISH

Roller Placement

1. Stand in front of the seat, facing the machine.
2. Position your right leg so that the lower roller is behind and slightly above your right ankle. The upper roller should be just above your right knee. Do not place the roller on your kneecap.
3. Brace yourself by holding the seat for support.
4. Slowly begin by lifting your lower leg up toward your buttocks. Your upper leg should stay pressed against the upper roller.
5. Pause briefly when the knee joint is at a 90-degree angle.
6. Slowly let the resistance pull your leg back to the start position. Control the weight all the way back down.
7. Keep your body straight and fixed: only your lower leg should be moving.
8. Complete the set and repeat with your other leg.

Legs

Lying Leg Curl

Muscles Worked: Hamstrings Group

START

FINISH

1. Lie face down on the bench and position your ankles behind the rollers. Make sure your knees are just off the edge of the bench.
2. Grip the handles for support.
3. Slowly begin to bring both legs upward until the knee joint is at 90-degrees.
4. Pause momentarily in the upper position, then slowly let your legs return to the start position, resisting all the way back down.
5. Your hips should stay in contact with the bench throughout the exercise. Do not use momentum to bring the weights up.
6. This exercise is slow and controlled throughout the range of motion.

Leg Abduction

Muscles Worked: Gluteus Medius, Gluteus Minimus

1. Attach the ankle strap around your right ankle and connect it to the low pulley.
2. Stand with your left side toward the machine, gripping the bench for support.
3. With your weight supported on the left leg, the right foot should be about an inch off the ground and directly in front of your left foot.
4. Slowly pull your right leg away from your body in a sideways motion.
5. Your body should be completely vertical: don't bend forward at the hips.
6. Pause briefly in the upper position, then slowly lower back to the start position.
7. After completing a set, repeat with the opposite leg.

Leg Adduction

Muscles Worked

Adductor Magnus, Adductor Longus, Gracilis, Pectineus, Adductor Brevis

1. Attach the ankle strap around your left ankle and connect it to the low pulley.
2. Stand slightly away from the machine, with your left side toward the machine, gripping the bench for support.
3. With your weight supported on the right leg, the left foot should be as close to the low pulley as possible while still maintaining tension on the cable.
4. Slowly pull your left leg toward your body in a sideways motion, crossing it in front of your right leg.
5. Your body should be completely vertical: don't bend forward at the hips.
6. Pause briefly in the upper position, then slowly lower back to the start position.
7. After completing a set, repeat with the opposite leg.

Standing Calf Raise

Muscles Worked: Gastrocnemius

1. The resistance source for this exercise can vary depending on the type of machine. Some will have shoulder supports and others will use a belt with an attachment to the low pulley. Regardless of the resistance source, the basic exercise is the same.
2. Stand upright, knees slightly bent and your feet spaced about 8 inches apart.
3. Position yourself so that the balls of your feet are supported on the platform and your heels as well as your arches are hanging over the edge.
4. Slowly rise up on your tiptoes into the start position, keeping your back straight.
5. Pause briefly in the upper position, then slowly lower yourself down until you feel a stretch in your calves.
6. Pause briefly in the lower position, then slowly push back to the start position.

Seated Calf Extension

Muscles Worked: Gastrocnemius

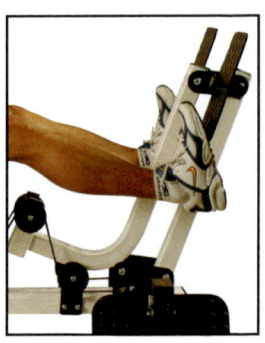

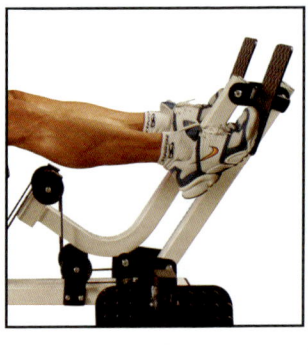

1. Place the balls of your feet on the lower part of the foot pad so your heels as well as your arches are hanging over the bottom edge.
2. Push the foot pad away from you until your legs are almost completely extended. Always keep a slight bend in the knees.
3. Slowly push the foot pad further with your toes, keeping your legs and upper body fixed. (Same motion as standing on tiptoes.)
4. Pause, then slowly release the tension, letting your toes be pushed back until you feel a slight stretch in your calf muscles.
5. Pause then push back to extended position.

Other Products by Productive Fitness Products Inc.

Home Gym Exercises | **63**

The Great Handbook Series

The Great Handbook series are a wonderful addition to your exercise library. These books have all the different exercises you need for working your whole body. In addition, they discuss how to set up a program, how to stretch, how to stay motivated, and safety tips. All have 64 pages of exercises using popular pieces of fitness equipment. The books, sold separately, are written and edited by experts in a clear and concise manner, with step-by-step instructions and full color photos for all exercises.

 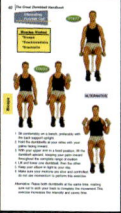 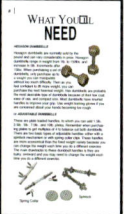   

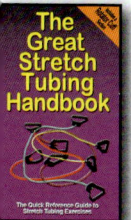  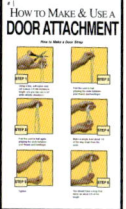   

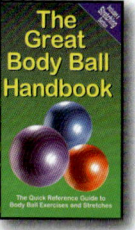     

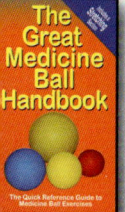     

Body Ball Training Poster Pack
Dumbbell Training Poster Pack
Stretch Tubing Training Poster Pack

Poster Packs Include Four or Five Full-Color
12" x 18" Posters Sold as Sets Only

The Ultimate Weight Training Journal

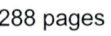

More than a <u>one-year personal fitness diary</u>, The Ultimate Weight Training Journal discusses basic nutrition, aerobics, and strength training. But best of all, this book shows you how these three tools can best be used in attaining a better physique, better health and more strength.

288 pages

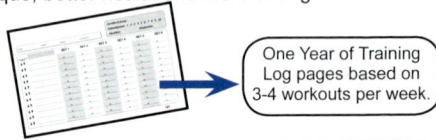

One Year of Training Log pages based on 3-4 workouts per week.

Fitness Poster Series
Full-Color **24" x 36"** Posters sold individually laminated, paper or framed

Body Ball-Core

Body Ball Upper/Lower Body

Dumbbell Shoulders/Arms

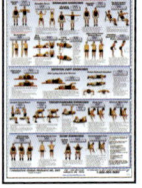

Dumbbell-Lower Body/Core/Chest/Back

Stretching Upper Body

Stretching Lower Body

Female Muscle Diagram

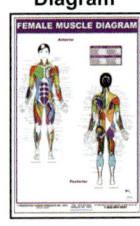

Male Muscle Diagram

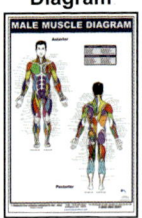

Fitness Heart Rate Chart

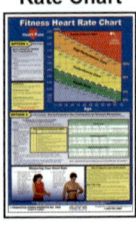

Home Gym Exercises